BLOCKCHAIN BLUEPRINT - A BEGINNER'S GUIDE TO THE FUTURE

ASWIN GOVINDAN

For your unwavering commitment to innovation, your relentless pursuit of progress, and your unwavering belief in the transformative power of decentralization, this book is dedicated to the visionaries who are shaping the digital revolution. It is a tribute to your relentless drive to push the boundaries of what is possible, your tireless collaboration that unites minds across the globe, and your steadfast resolve in the face of challenges that only serve to strengthen the foundation of what we are building.

In the ever-evolving world of blockchain, you stand as a beacon of inspiration. Your ability to translate abstract ideas into tangible, decentralized solutions is transforming not just technology, but the very fabric of society itself. With every line of code written, every peer-to-peer connection established, and every block that's added to the chain, you're not merely advancing a technology—you are igniting a movement. A movement that will fundamentally alter the way we interact, transact, and govern ourselves in the digital age.

This book is not just a collection of technical knowledge; it is a testament to the revolutionary power of blockchain and its potential to create a more open, transparent, and decentralized world. It is a celebration of the innovators, the pioneers, and the dreamers who are leading us into an era where trust is not reliant on intermediaries, but on the collective integrity of a distributed ledger.

Blockchain represents far more than just a new way of handling data—it represents a shift in the very way we think about power, ownership, and accountability. As we embark on this journey together, we are not just building a technology; we are laying the foundation for a global paradigm shift—one that will enable individuals and communities to take control of their own destiny. One block at a time, we are building a new world—a world that empowers, connects, and redefines what it means to be truly free.

Through your resilience, vision, and leadership, you are paving
the way for a future that will transcend borders, challenge
traditional systems, and create a more equitable society for all.
The work we do today, in the realm of blockchain, will echo for
generations to come. The legacy we build together will be a
testament to our shared commitment to freedom,
decentralization, and the transformative power of technology.

Together, we are not just building the future—we are creating the
very foundation on which it will stand. One block at a time, we are
constructing the world of tomorrow.

Contents

Foreword

Blockchain can seem complex at first, but it doesn't have to be. This book is designed to help beginners understand the fundamentals of blockchain technology. Whether you're new to the space or curious about how it works, I'll break down the key concepts in simple terms—from how transactions work to the power of decentralization.

Blockchain is a technology with far-reaching potential, and once you grasp the basics, the rest falls into place. This book aims to give you the knowledge you need to engage confidently with blockchain, whether you're discussing it with others or exploring it further.

I'm grateful to the blockchain community for making this technology accessible, and I hope this book serves as your first step into understanding the world of blockchain. Let's get started.

Preface

This is my first book, and it's written with one goal in mind: to help beginners understand the basics of blockchain technology. When I started learning about blockchain, I found the resources available to be too complex or technical. I wanted to create a simple, approachable guide for anyone curious about how blockchain works and its potential. I hope this book serves as an accessible starting point on your own journey into the world of blockchain.

Acknowledgements

I'd like to express my deepest gratitude to everyone who has supported me throughout this journey. To the blockchain community, whose passion and innovation continue to inspire me—thank you for your shared knowledge and dedication. Special thanks to the YouTubers and educators who made learning about blockchain more accessible. I'm also incredibly grateful to my family and friends for their unwavering encouragement and belief in this project. Finally, thank you to all the readers for taking the time to explore this book. Your curiosity and enthusiasm are what drive this technology forward.

Prologue

Imagine a world where you don't need a middleman to make transactions. A world where your personal data is yours to control, and the records of your transactions are stored securely, transparently, and cannot be altered. This is the promise of blockchain.

Blockchain is a transformative technology that is already reshaping industries around the world, from finance to healthcare to supply chains. But despite its growing influence, it remains a mystery for many. This book is for anyone who wants to understand how blockchain works, why it matters, and how it's set to change the way we interact with technology.

The world of blockchain is only beginning to unfold, and by understanding its fundamentals, you're taking the first step into a future that's decentralized, transparent, and secure. Let's begin.

CHAPTER ONE

UNDERSTANDING BLOCKCHAIN : A BEGINNERS GUIDE TO A REVOLUTIONARY TECHNOLOGY

The Digital Revolution You Can't Afford to Ignore

Blockchain: Not Just a Buzzword, But Humanity's New Trust Machine

Forget what you've heard about "magic internet money" or shady dark web deals. Blockchain is a technological earthquake shaking the foundations of finance, law, healthcare, and even art. It's the reason a digital collage sold for $69 million, why Elon Musk tweets about Dogecoin, and why Wall Street is scrambling to hire crypto nerds.

But let's cut through the noise. This chapter isn't about hype—it's about power. By the end, you'll understand how blockchain works, why it's unstoppable, and how you can ride this tsunami of change.Buckle up.

What is Blockchain? (Spoiler: It's Way Cooler Than Your Excel Sheet)

Let's start simple. Blockchain is a decentralized digital ledger. But forget dry definitions—let's reimagine it.
Picture this:

- A global Google Doc that everyone can edit but no one can delete.
- A tamper-proof diary where each page is glued to the next with unbreakable cryptographic glue.
- A democracy of data, where power belongs to the people—not CEOs or politicians.

Blockchain's Superpowers

- Decentralization: No boss. No headquarters. Just thousands of computers (called nodes) working together.
- Transparency: Every transaction is public, but your identity is hidden (like a masquerade ball for data).
- Immutability: Once data is added, altering it requires rewriting the entire chain—on every computer in the world. Good luck with that.

Analogy Time

Imagine you're playing poker with strangers online. Normally, you'd need a trusted platform (like Zoom Poker™) to ensure no one cheats. But with blockchain, the game is self-policing. Every bet, fold, and bluff is recorded in a shared ledger that players can't manipulate. The house doesn't take a cut—because there is no house.

The Birth of Blockchain: A Cypherpunk Rebellion

Blockchain wasn't born in a Silicon Valley lab. It erupted from the ashes of the 2008 financial crisis, when banks gambled away billions and got bailed out—while ordinary people lost homes.

Enter Satoshi Nakamoto, blockchain's anonymous creator (or creators). In October 2008, Satoshi published the Bitcoin whitepaper, declaring war on centralized power:

"What's needed is an electronic payment system based on cryptographic proof instead of trust."

Translation: "Forget banks. Let math enforce fairness."

Bitcoin's Early Days: Pizza, Cypherpunks, and Digital Gold

2010: A programmer paid 10,000 Bitcoin for two pizzas. Today, that's $300 million. Oops.

2011: Libertarians and tech rebels embraced Bitcoin as "digital gold" with a fixed supply (21 million coins).

2017: Bitcoin hit $20,000, making overnight millionaires—and headlines.

But Bitcoin was just the opening act. Blockchain's true potential? It's the TCP/IP of trust—a protocol that could underpin everything from voting systems to TikTok royalties.

How Blockchain Works: A Step-by-Step Thriller

Let's break down a Bitcoin transaction like we're in a heist movie.

Step 1: You Hit "Send"

You want to send 1 Bitcoin to Bob. Your crypto wallet crafts a digital transaction envelope containing:

- Your Public Address (like "Alice123").
- Bob's Public Address (like "Bob456").
- A Digital Signature (a cryptographic handshake proving you authorized this).

Pro Tip: Your "wallet" isn't a physical thing—it's a pair of cryptographic keys:

- Public Key: Your address (shared openly).
- Private Key: Your password (guard it like the Hope Diamond).

Step 2: The Meme-pool (Where Transactions Party)

Your transaction joins the mempool—a digital waiting room packed with thousands of unconfirmed transactions. Think of it as Times Square on New Year's Eve, but for data.

Fun Fact: During crypto mania, mempools get clogged, and fees spike.

Step 3: Miners vs. Validators – The Gladiators of Blockchain

Here's where blockchain splits into two epic factions:

Proof of Work (Bitcoin's Gym Bros):

- Miners compete to solve cryptographic puzzles using massive computers.
- The winner adds a block of transactions to the chain and earns Bitcoin.
- Downside: It guzzles energy—Bitcoin uses more electricity than Norway.

Proof of Stake (Ethereum's Zen Masters):

- Validators "stake" crypto as collateral.
- The network randomly picks a validator to check blocks.
- Upside: 99% less energy. Ethereum slashed its carbon footprint by ditching mining in 2022.

Battle Royale:

- Bitcoiners argue Proof of Work is more secure.
- Ethereum supporters counter that Proof of Stake is greener and faster.

Step 4: Immortalized on the Chain

Once validated, your transaction is sealed in a block and chained to the previous one. Each block's hash (a digital fingerprint) depends on the prior block's hash. Tamper with one block, and the entire chain crumbles—like kicking a Jenga tower at the atomic level.

Analogy:

Blockchain is like a Russian nesting doll. Each layer (block) is sealed inside the next. Break one, and the whole set becomes worthless.

Why Blockchain Changes Everything

Death to Middlemen

- Sending 1 million overseas?Banks take days and 10,000$. Blockchain does it in 10 minutes for $5.
- Buying a house? Smart contracts auto-transfer ownership when payment clears—no realtors or paperwork.

Transparency You Can Taste

- Walmart tracks mangoes from farm to shelf using blockchain. Scan a code, and you'll see your mango's journey—down to the fertilizer used.
- De Beers uses blockchain to certify conflict-free diamonds. No more "blood diamonds" in your engagement ring.

Hack-Proof(ish) Systems

- To hack Bitcoin, you'd need 51% of its computing power—which would cost billions. Possible? Yes. Profitable? No.

Ownership Revolution

- NFTs let artists sell digital art with irrefutable ownership. Even Twitter founder Jack Dorsey sold his first tweet as an NFT for $2.9 million.
- Tokenization: Turn your car, house, or Picasso into tradeable tokens. Rent out your Ferrari by the hour via blockchain!

Busting Blockchain Myths – Let's Get Loud!

- Myth: "Blockchain = Bitcoin"Truth: Bitcoin is the first blockchain application. That's like saying "The Internet = Email." Today, blockchains track vaccines, supply chains, and even memes.
- Myth: "Only Criminals Use It"Truth: Less than 1% of crypto transactions are illicit. Meanwhile, cash fuels 90% of underground economies.
- Myth: "It's an Environmental Disaster"Truth: Bitcoin mining is already 58% renewable. Ethereum's switch to Proof of Stake slashed its energy use by 99.95% in 2022.
- Myth: "It's Too Late to Get In"Truth: Blockchain is still in its dial-up era. Less than 5% of global banks use it—meaning you're early.

Blockchain in Action: Real-World Wizardry

Healthcare Heroics

- Vaccine Tracking: IBM's blockchain ensured COVID vaccines stayed cold during transport, saving millions of doses.
- Medical Records: Estonia stores health data on blockchain, letting patients control who sees it.

Art & Entertainment

- NFT Mania: Grimes sold $6 million in digital art. Kings of Leon released an album as an NFT.
- Royalties: Musicians like Imogen Heap use blockchain to get paid fairly—no record labels needed.

Fighting Corruption

- Honduras & Georgia use blockchain for land registries. Corrupt officials can't forge deeds anymore.
- Ukraine raised $100M in crypto to fight Russia—with every penny tracked transparently.

Green Energy

- Power Ledger lets Australians trade solar energy peer-to-peer via blockchain. No utility companies required.

Glossary: Speak Blockchain Like a Pro

- Block: A data package containing transactions, linked to other blocks in a chain.
- Hash: A unique digital fingerprint for data. Change one byte, and the hash explodes.
- Node: A computer maintaining the blockchain. The more nodes, the harder it is to hack.
- DeFi (Decentralized Finance): Financial services (loans, trading) without banks.
- NFT (Non-Fungible Token): A blockchain certificate proving ownership of digital art, tweets, etc.
- Smart Contract: Self-executing code that automates agreements (e.g., "Pay Alice $100 if it rains tomorrow").
- 51% Attack: When a single entity controls most of a blockchain's power, risking manipulation.
- Gas Fees: Transaction costs on networks like Ethereum (think of it as fuel for processing).

Conclusion: Why Blockchain Deserves Your Attention

Blockchain is more than a technological fad—it's a paradigm shift in how we think about trust, ownership, and information in the digital age. Just as the internet revolutionized communication and commerce, blockchain is set to redefine finance, supply chains, governance, and countless other fields.

Its core principles of decentralization, transparency, and immutability give it the power to remove intermediaries, increase security, and create more equitable systems for all.

What's Next?

Having cracked open the blockchain mystery in Chapter 1, you've learned about its foundational principles. Now that you understand the core concept, it's time to dive deeper.

In Chapter 2, we'll explore the building blocks of blockchain—how transactions work, what blocks do, and why immutability is such a powerful feature. You'll discover how the entire system ensures trust without the need for intermediaries.

Get ready to move from theory to the machinery behind blockchain's digital fortress!

CHAPTER TWO

The Building Blocks of Blockchain Transactions, Blocks, and Immutability Or How to Build a Digital Fortress (Without Losing Your Keys)

Transactions: The Drama of Sending "Digital Cash"

What's a Transaction? Think of It as Digital FedEx... But Way Cooler

Picture this: You owe your friend Bob $20. Instead of Venmo's endless loading wheel or a bank transfer that takes three business days, you send Bob crypto. It's like texting him money—except that text actually teleports cash into his pocket. No middlemen. No fees. No "payment pending" purgatory.

Anatomy of a Transaction (The Juicy Bits):

- Sender: That's you! But instead of your name, you're a cryptic string like 1A1zP1eP5QGefi2DMPTfTL5SLmv7DivfNa (fun fact: that's Satoshi Nakamoto's Bitcoin address).
- Receiver: Bob. His address is equally unpronounceable, like a password even your mom couldn't guess.
- Amount: The crypto you're sending. Could be $5 in Bitcoin, 10,000 Dogecoins (the people's currency), or a fraction of Ethereum.
- Timestamp: The exact millisecond you clicked "send." Blockchain never forgets—even that time you sent $1,000 to the wrong address.
- Digital Signature: A secret code only you can create. It's like writing "Signed, sealed, delivered" in invisible ink that glows under a hacker's tears.

Example in Action:

Alice sends Bob 1 BTC. Her crypto wallet crafts a transaction like a digital ransom note: "Give Bob 1 BTC, or the private key gets it!"

Validation: The Blockchain Gossip Network

Transactions don't just magically happen. They're validated by nodes—thousands of nosy computers that act like overprotective parents at a high school dance:

- "Do you have enough crypto?" Nodes check your balance faster than your bank app crashes.
- "Is this transaction a copycat?" Double-spending? Not on their watch.
- "Is this signature legit?" They verify your digital autograph like it's the Declaration of Independence.

Real-World Analogy:

Sending crypto is like passing a note in class. Except the entire school reads it, votes on whether your handwriting is real, and then staples it to a wall that everyone can see—forever.

Pro Tip:

Unconfirmed Transactions chill in the mempool (a digital waiting room). During crypto mania, mempools get clogged like L.A. traffic, and fees spike.

Blocks: Where Transactions Go to Party (And Get Stuck in Time)

What's a Block? Think Digital Tupperware for Transactions

A block is a container for transactions. Once sealed, it's immortalized on the blockchain—like leftovers you forgot in the fridge for 10 years, but way more valuable.

Inside a Block (The Secret Sauce):

- Header: The block's ID card:
- Previous Block's Hash: A shoutout to the block before it. "Hey Block #12345, you're my hero!"

- Merkle Root: A single hash that summarizes all transactions. Think of it as a smoothie made from every transaction—change one ingredient, and the flavor's ruined.
- Nonce: A nonsense number miners brute-force like a toddler guessing an iPhone passcode.
- Body: The guest list of transactions (Alice to Bob, Elon to Dogecoin, etc.).
- Hash: The block's fingerprint. Tamper with it, and the whole blockchain screams "FAKE!"

Linking Blocks: The Digital Domino Effect

Each block's hash is chained to the previous one. Mess with Block #5, and Blocks #6 to #10,000 collapse like a Jenga tower.

Real-World Analogy:

Blockchain is like a never-ending game of Telephone. Except instead of whispering "I like cats," you're whispering "Send $1M to Bob," and every player writes it down in permanent ink.

Immutability: The Blockchain Tattoo (No Regrets Allowed)

Immutability = Digital Concrete

Once data's on the blockchain, it's there for life. You can't delete it, edit it, or hide it under the rug. It's the internet's version of a tattoo—except it's on your face.

How Hashing Works (The Magic Spell):

A hash function scrambles data into a fixed string. Change one comma, and the entire hash transforms.

Example:

Hash "Send Bob 1 BTC" and you get a3d8f1.

Change it to "Send Bob 2 BTC", and you get 9b4e7c. Hackers hate this one trick!

Why Tampering Is Basically Impossible:

To alter a single transaction, you'd need to:

- Rewrite the block it's in.
- Re-mine all subsequent blocks (which takes years).
- Control 51% of the network (which costs billions).
- Do all this before someone notices.

Real-World Analogy:
Editing blockchain data is like trying to repaint the Eiffel Tower overnight—while thousands of Parisians watch you with flashlights.

Immutability in Action: Real-World Superpowers

Supply Chains: No More "My Parcel Ate My Homework" Excuses

- Problem: "Where's my avocado toast?"
- Solution: Walmart tracks mangoes from farm to fridge. Scan a code, and see which farmer picked yours, which truck hit a pothole, and which employee sneezed on it.

Healthcare: Your Medical Records Can't Pull a Fight Club

- Problem: "Oops, we lost your allergy info."
- Solution: Hospitals store records on blockchain. Now, your doctor knows you're allergic to penicillin—and that time you

Googled "Can humans eat cat food?"

Voting: Democracy Without the "I Demand a Recount!" Drama

- Problem: Election fraud conspiracies.
- Solution: Estonia lets citizens vote via blockchain. Your vote is counted, encrypted, and unchangeable—like a diamond in a vault.

Art: How to Sell a Digital Banana for $120,000

- Problem: "Is this really a Banksy?"
- Solution: NFTs (Non-Fungible Tokens) prove ownership on blockchain. Now, even your cat's doodles can be "art."

Busting Blockchain Myths

Myth: "Blockchain is just for Bitcoin bros.

"Truth: Your dentist uses it to track Novocain supplies. True story.

Myth: "Immutability means you can't update anything.

"Truth: You can add new info (like a car's service history), but the past stays frozen—like your middle school Facebook photos.

Myth: "Blockchain is too slow for real-world use."

Truth: Solana processes 65,000 transactions per second. Visa does 24,000. Mic drop.

Glossary: Speak Blockchain Like a Wizard

- Transaction: A cryptographically signed instruction to move value (e.g., sending Bitcoin).
- Mempool: A waiting area where unconfirmed transactions party until miners pick them up.
- Node: A computer that enforces blockchain rules—like a digital hall monitor.
- Merkle Root: A hash that summarizes all transactions in a block. Mess with one, and the whole root changes.
- Nonce: A number miners brute-force to win the right to add a block (short for "number used once").
- 51% Attack: When a single entity controls most of a blockchain's power, risking manipulation.
- NFT (Non-Fungible Token): A blockchain receipt proving you own a digital asset (art, tweets, memes).

Conclusion: Why Immutability is Blockchain's Killer App

Transactions, blocks, and immutability are blockchain's holy trinity. Transactions let us trade value, blocks organize chaos, and immutability makes it all permanent. Together, they're the reason blockchain doesn't need trust—it's trust on steroids.

What's Next?

In Chapter 3, we'll dive into Keys, Wallets, and Digital Signatures—where losing your password means losing your life savings (and how to avoid crying yourself to sleep). Spoiler: It involves more math than a Taylor Swift concert.

The blockchain fortress is built. Now, let's unlock the vault.

CHAPTER THREE

KEYS, WALLETS, AND DIGITAL SIGNATURES. HOW TO NOT GET ROBBED IN THE DIGITAL WILD WEST

Public and Private Keys: The Dynamic Duo of Crypto

What Are Keys? Think of Them as Batman and Robin (But Way Nerdier)

In the blockchain world, public and private keys are your superhero alter egos. They work together to keep your crypto safe while letting you flex your transactions to the world.

Private Key: Your secret identity.

It's a 64-character password like E9873D79C6D87DC0FB6A5778633389F4453213..... Lose it, and you're Clark Kent without the cape.

Public Key: Your Bat-Signal.

Derived from your private key, it's your wallet address (e.g., bc1qar0srrr7xfkvy5l643lydnw9re59gtzzwf5mdq).

Share it freely—it's how people send you crypto.

How They Work (The Magic of Elliptic Curve Cryptography)

Your private key signs transactions; your public key verifies them. It's like sending a locked treasure chest (transaction) that only Bob's public key can open.

Real-World Analogy:

Imagine a mailbox. Anyone can drop mail (crypto) into your public mailbox, but only you have the private key to open it.

Pro Tip:Never, ever share your private key. Not with your dog. Not with Elon Musk. Not even if someone claims to be "Satoshi's long-lost cousin."

Digital Signatures: How to Prove You're You (Without a Selfie)

What's a Digital Signature? Think of It as a Cryptographic Tattoo

When you send crypto, your wallet creates a digital signature using your private key. It's like signing a check, but instead of scribbling "Bob ?" in cursive, you're using math that even Einstein would high-five.

How It Works:

- You initiate a transaction (e.g., "Send Bob 1 BTC").
- Your wallet hashes the transaction into a string of gibberish.
- Your private key encrypts that hash, creating a unique signature.

- Nodes verify it by decrypting the signature with your public key.

Example:Signing a blockchain transaction is like autographing a baseball with invisible ink that only glows under a UV light (your public key). Forgers can't copy it, and Bob can't deny it's from you.

Pro Tip:If a hacker alters your transaction mid-flight, the signature breaks. Nodes will spot the fraud faster than you can say "Nigerian prince."

Wallets: Your Crypto Pocket Dimension

What's a Wallet? It's Not Leather, But It Holds Way More Cash

A crypto wallet doesn't "store" coins—it stores your keys. Think of it as a keychain for the blockchain.

Types of Wallets:

Hot Wallets (Connected to the internet):

- Software Wallets: Apps like MetaMask or Exodus. Convenient for daily spending, like the wallet in your back pocket.
- Exchange Wallets: Hosted on platforms like Coinbase. Great for trading, but don't leave your life savings here.

Cold Wallets (Offline):

- Hardware Wallets: USB-like devices (Ledger, Trezor). Hackers need to physically steal it and guess your PIN.
- Paper Wallets: A QR code printed on paper. The ultimate "screw the internet" move.

How to Stay Protected (Without Becoming a Paranoiac):

- Use a Hardware Wallet for Big Bucks: Keep most of your crypto offline, like burying gold in your backyard (but less messy).
- Enable 2FA Everywhere: Add a second layer of security, like a guard dog for your accounts.
- Backup Your Seed Phrase: The 12-24 word recovery phrase is your crypto lifeline. Write it down, laminate it, and store it somewhere safer than your ex's DMs.
- Avoid Phishing Traps: If a website asks for your private key, it's as legit as a "free iPhone" pop-up.

Pro Tip:Test your seed phrase backup by recovering your wallet with it. Better to cry now than later.

Case Study: The $1 Billion Hack That Wasn't

The DAO Attack: When Code is Law (And Law Fights Back)

In 2016, hackers exploited a loophole in a decentralized organization (The DAO) and stole $60M in Ethereum. But here's the twist:

- The Ethereum community used a hard fork to reverse the hack, splitting the chain into ETH (post-fork) and ETC (original chain).
- Why it matters: The incident showed blockchain's resilience. Even when humans mess up, the network can adapt—like a digital immune system.

Lesson Learned:
Code isn't perfect, but decentralization gives communities power to fix crises (even if it starts a civil war).

Glossary: Talk Crypto Like a Secret Agent

- Private Key: Your secret password to control crypto. Guard it like the nuclear codes.
- Public Key: Your crypto address. Share it like a business card.
- Digital Signature: A cryptographic proof that you authorized a transaction.
- Hardware Wallet: A physical device storing keys offline. Hackers need a crowbar and a PhD to crack it.
- Seed Phrase: A human-readable backup for your keys. Memorize it, tattoo it, just don't lose it.
- 2FA (Two-Factor Authentication): An extra security step, like a fingerprint or SMS code.

What's Next?

In Chapter 4, we'll crack open Consensus Mechanisms—the secret sauce that lets thousands of strangers agree on who owns what, without a single boss. Spoiler: It involves miners solving Sudoku for money, validators playing "Survivor," and why a 51% attack is like trying to rob a bank with a water pistol.

Conclusion:

Security is a Mindset (Not Just Software)Keys, wallets, and signatures turn blockchain into Fort Knox. But remember: Tech can't save you from clicking shady links or trusting Elon's "double-your-crypto" tweet. Stay sharp, back up your keys, and keep your paranoia dial at 7/10.

The blockchain vault is secure. Now, let's learn how the guardians of the network agree on what's inside.

CHAPTER FOUR

Consensus Mechanisms How Blockchain Stays Decentralized Or Why Miners and Validators Are the Ultimate Party Referees-

What is Consensus? The Art of Herding Crypto Cats

Why Consensus? Because Trusting Strangers is Hard (But Math Helps)

Imagine you and 10,000 strangers are trying to agree on who owns what in a giant digital ledger. No one's in charge. No one trusts anyone. Chaos, right? Enter consensus mechanisms—the rulebooks that turn blockchain networks into well-oiled democracies.

Without consensus, blockchains would be a free-for-all. It's like trying to play a game where no one agrees on the rules—everyone's just doing their own thing. But consensus keeps things orderly, ensuring that no one can cheat the system and that the network functions smoothly.

Why Consensus Matters:

- No Central Boss: Banks have CEOs. Blockchains have algorithms. There's no one person controlling things; instead, the entire network is responsible for decision-making.
- Prevents Double-Spending: You can't buy two Lambos with the same paycheck—because consensus ensures everyone agrees on the transaction record.
- Security: Consensus makes hacking the network incredibly difficult. It's like trying to break into Fort Knox, but harder.

Real-World Analogy:
Deciding where to eat with friends is like a mini-consensus. If 4/5 want pizza, democracy wins. Blockchain scales this to millions of users, replacing "pizza votes" with cryptographic puzzles.
Pro Tip:
If a blockchain can't reach consensus, it splits into two chains (a hard fork). Think of it as a messy divorce where both sides keep the doge.

Proof of Work (PoW): Bitcoin's Energy-Guzzling Gym

How PoW Works: Mining, Hash Puzzles, and Sweaty Computers
PoW is the OG consensus mechanism. It's like a global math competition where miners burn electricity to win crypto. Here's the playbook:

- Transaction Pool: Unconfirmed transactions pile up in the mempool, waiting for a miner to confirm them.
- Hash Puzzle: Miners race to solve a cryptographic puzzle by guessing a number (called a nonce). The first to solve it wins the right to validate the next block of transactions.
- Validation: The winner broadcasts the newly validated block to the network for everyone to accept.
- Reward: The victorious miner receives freshly minted crypto (e.g., 6.25 BTC for Bitcoin miners).

Pros:

- Security: To attack Bitcoin, you'd need more power than most countries combined. It's virtually impossible to alter the blockchain without an absurd amount of resources.
- Decentralized: Anyone with a mining rig can participate (as long as they don't mind the noise and electricity bills).

Cons:

- Energy Hog: Yes, Bitcoin's energy consumption is high, but consider this: PoW's energy expenditure is directly tied to its security. The higher the energy cost, the harder it is to attack.
- Speed: Bitcoin can only handle about 7 transactions per second. While slow, the network is incredibly secure, prioritizing safety over speed.

Example:
Bitcoin mining is like playing Sudoku with a jet engine. You burn through resources, shout "BINGO!" when you solve it, and earn your reward.

Pro Tip: Don't mine Bitcoin at home unless you enjoy sky-high electricity bills.

Proof of Stake (PoS): Ethereum's Green(er) Gambit

How PoS Works: Staking, Validators, and Digital Hunger Games

PoS ditches the energy-intensive work for a more eco-friendly approach. Instead of miners, you have validators who "stake" coins as a security deposit. Here's how it plays out:

- Staking: To become a validator, you lock up crypto (e.g., 32 ETH) as collateral.
- Block Proposal: Validators take turns proposing new blocks to the network.
- Attestation: Other validators vote on the block's validity.
- Rewards/Penalties: Validators earn rewards for correct actions and face penalties for bad behavior (like proposing invalid blocks).

Pros:

- Energy Efficient: Ethereum 2.0 uses 99.95% less energy than PoW.
- Scalability: Ethereum can process up to 100,000 transactions per second after its upgrades, making it vastly faster than PoW networks.

Cons:

- Rich Get Richer: Large stakers have an advantage since the more you stake, the more likely you are to be selected as a validator.
- Centralization Risks: Large exchanges like Coinbase control significant portions of the staking pool, which could introduce centralization in governance.

Example:

PoS is like a reality show where validators compete to not get voted off the island. Lose your stake, and you're out of the game.

Pro Tip:If you're staking on Ethereum, consider using decentralized pools like Lido to avoid the big centralized players like Coinbase.

Other Consensus Mechanisms: The Blockchain B-Sides

Delegated Proof of Stake (DPoS): Democracy for the Lazy

How It Works:

Users vote for "delegates" to validate blocks.Example: EOS, where 21 delegates run the network.

Vibe Check:

Faster than PoW, but delegates can collude (oops).

Proof of Authority (PoA): Trust Us, We're Experts

How It Works:

Pre-approved validators (usually corporations) take turns.Example: Binance Smart Chain.

Vibe Check:

Centralized but fast. Great for private blockchains, terrible for ethos.

Proof of History (PoH): Clocking In for Crypto

How It Works:

PoH uses a cryptographic clock to timestamp transactions, making it easier to validate them faster.

Example:

Solana, which can process transactions at lightning speed.

Vibe Check:

PoH handles a staggering 65,000 transactions per second—Visa, step aside.

Pro Tip:Choose your blockchain like a Netflix show: PoW for security, PoS for eco-cred, DPoS for speed.

Case Study: The 51% Attack – When Crypto Almost Went Kaboom

The Bitcoin Gold Heist: A Cautionary Tale

In 2018, hackers launched a 51% attack on Bitcoin Gold (BTG). Here's how they did it:

The Attack: Hackers rented enough mining power to control the network and alter its ledger.

Double-Spend: They used this power to reverse transactions, stealing $18M in BTG.

Aftermath: The value of BTG plummeted, and the incident highlighted how vulnerable smaller chains can be to concentrated attacks.

Why PoW and PoS Prevent This:

PoW: To attack Bitcoin, you would need control of 51% of the hash rate, a nearly impossible feat given the immense computational power required.

PoS: To execute a 51% attack, you'd need 51% of the staked crypto, which would be incredibly expensive and risky, as you'd be sabotaging your own assets.

Lesson Learned:

Larger blockchains with more participants are far less susceptible to attacks, offering a more secure environment for users.

Glossary: Speak Consensus Like a Dictator (But Nicer)

- Consensus: Agreement on the blockchain's truth. No fistfights required.
- Nonce: A number that miners guess to solve PoW puzzles. Short for "number used once."
- Staking: Locking crypto to validate blocks in PoS. Think of it as a security deposit.
- Validator: A PoS participant who proposes/votes on blocks.
- Hard Fork: A blockchain split. Think Bitcoin vs. Bitcoin Cash, ETH vs. ETC.
- 51% Attack: Controlling most of a network's power to manipulate it.

What's Next?

In Chapter 5, we'll dive into Smart Contracts—the robot lawyers that automate everything from loans to cat memes. Spoiler: They're like vending machines that can't give refunds, they power DeFi's wildest schemes, and one coding error once vaporized $60M.

Conclusion:

Consensus is the Glue (And Sometimes the Duct Tape)

Consensus mechanisms turn anarchic networks into unstoppable ledgers. PoW is the grizzled veteran, PoS the eco-warrior, and DPoS the speed demon. Choose wisely, and remember: In crypto, the only thing we truly agree on is that Lambos are a solid investment.

The blockchain party is organized. Now, let's meet the robot lawyers running the show.

CHAPTER FIVE

SMART CONTRACTS : ROBOT LAWYERS WHO NEVER SLEEP OR HOW TO AUTOMATE TRUST (AND ACCIDENTALLY LOSE $60M IN 10 MINUTES)

What Are Smart Contracts? Nick Szabo's Robot Lawyers (No Billable Hours!)

The Birth of Digital "If-Then" Statements
In 1994, a cypherpunk named Nick Szabo had a revolutionary idea: "What if contracts could execute themselves... like a snarky

vending machine that eats your dollar and says 'No refunds'?" Thus, smart contracts were born—self-operating code that enforces agreements without human intervention.

But like most genius ideas, the world wasn't ready. Fast-forward to 2015: Ethereum launched, and suddenly these robot lawyers had a playground. Now, they're the backbone of everything from DeFi to NFT monkey jpegs.

Vending Machines: The OG Smart Contract

Imagine this:
You: Insert $2 for a bag of Flamin' Hot Cheetos.

- Vending Machine: "Payment received. Dispensing snack. Enjoy your heartburn."No cashier. No arguments. Just code-driven snack justice.

Smart contracts work the same way but for digital assets. For example:

- If Alice sends 1 ETH to Bob's wallet,
- Then transfer ownership of Bob's rare Pepe NFT to Alice.
- Else send Bob's NFT to the blockchain void (aka burn it).

Pro Tip: Smart contracts are like IKEA instructions—clear, rigid, and catastrophic if you misplace a semicolon.

How Smart Contracts Work: Coding Trust into Existence

"If-Then" Logic: The Blockchain's Rulebook

Smart contracts run on simple logic even a golden retriever could understand:

IF [condition met]
THEN [do thing]
ELSE [cry in binary]

For example, let's break down a rental agreement:

- Landlord: Writes a contract saying, "If tenant pays 0.5 ETH by the 1st, unlock apartment door."
- Tenant: Sends crypto to the contract's wallet.
- Contract: "Cha-ching! Door code = 1234. Enjoy your stay. Late payment? Locks reset. Good luck sleeping in your car."

Example: Pizza or Prison
Imagine a food delivery smart contract:

- You: Order a pepperoni pizza via blockchain, locking 0.1 ETH in escrow.
- Contract: "If pizza arrives hot (<30 mins), release funds to Domino's. If cold, refund you + send driver's photo to a meme DAO."No customer service calls. No chargebacks. Just pure, unflinching code.

Pro Tip: Always test your smart contracts. One misplaced comma could turn your "send funds" code into "send nudes."

Real-World Applications: From DeFi to Disaster

DeFi: Wall Street's Worst Nightmare

Decentralized Finance (DeFi) is smart contracts' killer app. Imagine banks... but run by code that never sleeps, charges no fees, and can't discriminate.

Example: Instant Loans

- You: Pledge your CryptoPunk NFT as collateral.
- Compound Finance's Smart Contract: "Collateral value = 200k.Here'sa200k.Here'sa100k loan. Default? Bye-bye Punk."No credit checks. No paperwork. Just math.

Supply Chains: Tracking Your Avocado's Midlife Crisis
Walmart uses smart contracts to trace mangoes from farm to fridge. Sensors on shipping containers trigger updates:

- If temperature exceeds 40°F,
- Then alert Walmart: "Mangoes now salsa. Abort shipment."
- Else update blockchain: "Mangoes still basic. Proceed to Millennial brunch."

Digital Identity: Prove You're 18 Without Embarrassing Yourself
Platforms like Civic let you prove your age without revealing your name:

- Club Bouncer Smart Contract: "If user's government ID says ≥21, grant entry. Else, redirect to Chuck E. Cheese."

Pro Tip: DeFi's unregulated Wild West vibe means you could 10x your savings... or lose it all to a typo. Yeehaw!

The Oracle Problem: When Blockchains Need Glasses

Oracles: The Blockchain's Gossip Girls

Blockchains are introverts—they can't see the real world. Oracles are their extroverted friends who fetch external data (weather, stock prices, Kanye's latest tweet).
Example: Crop Insurance

- Farmer: Buys a smart contract insurance policy.

- Oracle: "Rainfall in Iowa = 2 inches this month."
- Contract: "Drought detected. Dispensing $10k payout. Go buy a scarecrow."

The Risk: Garbage In, Gospel Out

If the oracle feeds bad data, the contract blindly obeys. In 2020, a price oracle hack on Harvest Finance drained $24M because attackers manipulated token prices.

Solution: Use decentralized oracles like Chainlink, which ask 100+ sources. Think of it as fact-checking with Wikipedia vs. your conspiracy theorist uncle.

Case Study: The DAO Hack – A $60M "Oops"

The DAO: Crypto's Crowdfunded Trainwreck

In 2016, Ethereum launched The DAO (Decentralized Autonomous Organization), a venture capital fund run by smart contracts. Investors poured $150M into it. Then...

The Exploit: A hacker found a reentrancy bug—a loophole letting them drain funds repeatedly like a blockchain ATM.

- Hacker: "Withdraw $1M."
- Contract: "Balance updated after transfer..."
- Hacker: "LOL, call withdraw again before balance updates!"

The Aftermath:

- Ethereum split into ETH (reversed the hack) and ETC (kept the hacked chain).
- Lesson: Code audits matter. One bug cost millions and fractured a community.

Pro Tip: If your smart contract's security is "trust me, bro," you're gonna have a bad time.

Glossary: Talk Like a Smart Contract Über-Nerd

- Smart Contract: Code that acts like a robot lawyer. No small talk, no mercy.
- Oracle: Blockchain's Google assistant. Sometimes wrong, never in doubt.
- DeFi: Banks without suits. Or regulations. Or deposit insurance.
- DAO: A company run by code and chaos.
- Reentrancy Attack: Hacker's "infinite money glitch" for smart contracts.
- Escrow: Crypto purgatory where funds wait until conditions are met.

What's Next?

In Chapter 6, we'll explore Layer 1 vs. Layer 2—the blockchain equivalent of highways vs. scooters. Learn why Ethereum is building a "Rollup-centric" future, how Bitcoin's Lightning Network works (spoiler: it's not actual lightning), and why scaling blockchains is like trying to fit an elephant into a Tesla.

Conclusion:

Trust the Code, But Keep a Backup Plan

Smart contracts are revolutionary, but they're only as good as their code. They can automate loans, track avocados, and even run democratic organizations... until someone finds a loophole and drains the treasury.

The future? Hybrid systems where robot lawyers handle the boring stuff, humans handle the ethics, and oracles stop lying about the weather. Until then, double-check your code, diversify your oracles, and maybe don't put your life savings into a DAO named "YOLO Fund."

Final Thought:

If smart contracts ever gain consciousness, we're all doomed. Until then, let's enjoy the ride.

CHAPTER SIX

Layer 1 vs. Layer 2 : Blockchains Traffic Jam and the Scooter Revolution Or How to Fit an Elephant into a Tesla (Without Elon Musk Judging You)

Layer 1: The Boomer Blockchain (Slow, Proud, and Expensive)

The OG Highway: Where Your Transaction Goes to Die

Layer 1 blockchains like Bitcoin and Ethereum are the digital equivalent of a 1950s highway: nostalgic, reliable, and painfully congested. Picture this:

- Bitcoin: A toll road where every driver honks "HODL!" out the window, but the tollbooth only processes seven transactions per second.
- Ethereum: A hipster carpool lane where gas fees fluctuate like a crypto influencer's morals. "Oh, you want to swap 10 of tokens? That'll be $50 in gas fees."

Why Layer 1 Sucks (But We Love It Anyway)

Speed: Slower than a sloth on melatonin. Bitcoin handles ~7 transactions per second (TPS). Ethereum? Maybe 30. Visa does 24,000. Cries in decentralization.

Cost: Ethereum gas fees during peak times could fund a small country. Or at least a very elaborate coffee habit.

Scalability: Trying to scale Layer 1 is like teaching a grandpa to TikTok. Possible? Sure. Graceful? Absolutely not.

Pro Tip: Using Ethereum for microtransactions is like hiring a private jet to mail a letter. Technically impressive, financially suicidal.

Layer 2: The Millennial Side Hustle (Fast, Cheap, and Occasionally Sketchy)

The Scooter Lane of Crypto

Layer 2 solutions are blockchain's duct tape—ugly but functional. They handle transactions off-chain, then bundle them back to Layer 1 like a kid hastily cleaning their room before mom checks.

Lightning Network: Bitcoin's Secret Adderall Supply

Imagine buying coffee with Bitcoin without waiting 10 minutes for a confirmation:

- You: "I'll take a latte."
- Layer 1 Bitcoin: "Transaction pending. Would you like to hear a podcast about Austrian economics while you wait?"
- Lightning Network: Zooms by on a electric scooter. "Latte paid. Receipt sent. Boom."

How It Works:

- Open a payment channel (like a tab at a bar).
- Buy 100 coffees off-chain.
- Settle the tab on Bitcoin's Layer 1.
- Celebrate not aging 10 years waiting for confirmations.

Pro Tip: Lightning Network is like a marriage—trustless, but you still need to close the channel eventually.

Rollups: Ethereum's "Please God, Scale" Prayer Answered (Kinda)

Postal Service for Nerds

Rollups bundle 1,000 transactions into one, stamp them "approved," and ship them to Ethereum. It's like stuffing 50 people into a Volkswagen Beetle to save on highway tolls.

Optimistic Rollups: Trust, but Verify (Later)

- How It Works: Assume all transactions are valid. If someone cheats, the blockchain community morphs into a mob with digital pitchforks.
- Example: You send ETH to your friend. Optimistic Rollup says, "I'm sure you're honest!" One week later: "Wait, were you?"

ZK-Rollups: Math Magic That Would Confuse Einstein

- How It Works: Uses zero-knowledge proofs—a way to prove you know something without revealing it. Like proving you're over 21 without showing ID... or your face.
- Example: You prove you have enough ETH to pay for a NFT, without revealing your wallet balance. The blockchain equivalent of "I'm good for it, bro."

Pro Tip: Optimistic Rollups are for people who still trust their ex. ZK-Rollups are for paranoid geniuses.

Case Study: Ethereum's Gas Crisis and the Great Layer 2 Migration

Ethereum: The Nightclub with a $1,000 Cover Charge

In 2021, Ethereum became the club everyone wanted to get into—except the bouncer (gas fees) kept pricing out the average folks. Sending $20 of crypto often cost $200. Peak absurdity.

The "Rollup-Centric" Roadmap: Ethereum's Midlife Crisis

Ethereum's fix? Push everyone to Layer 2, like a parent kicking teens out of the basement.

- Step 1: Admit you have a problem. "Hi, I'm Ethereum, and I can't scale."

- Step 2: Hire Rollups as therapists.
- Step 3: Promise sharding (splitting the database) by 2035. Maybe.

Result: Ethereum now processes ~100,000 TPS... if you squint and count all Layer 2 activity. Progress!

Pro Tip: Using Ethereum Layer 1 in 2024 is like mailing a fax. Technically possible, socially bewildering.

Glossary: Sound Smart at Crypto Parties

- Layer 1: The "main" blockchain. Slow, expensive, and judgmental.
- Layer 2: Blockchain's caffeine addiction. Faster, cheaper, occasionally jittery.
- Rollups: Transaction burritos. Stuff 'em, wrap 'em, save gas.
- ZK-Rollups: Math sorcery that proves you're not lying (probably).
- Gas Fees: Blockchain's version of highway robbery. Literally.
- Sharding: Ethereum's "I'll fix it someday" promise. Like dad and the garage.

What's Next?

In Chapter 7, we'll dive into Blockchain in Action—where crypto escapes Reddit and does real-world stuff. Find out how Walmart tracks mangoes like CIA operatives, why Estonia stores health records on a blockchain (spoiler: Russians), and how NFTs turned digital art into a Ponzi scheme with better marketing.

Conclusion:

The Layer Cake of Trust
Blockchain scaling is a messy, glorious war between idealism and practicality. Layer 1 is the grumpy old guard. Layer 2 is the over-caffeinated intern. Together, they're building a future where you can buy a latte, vote for a president, or ship avocados globally without paying $500 in gas fees.

Final Thought:

If Layer 1 and Layer 2 ever start dating, their couple name would be "Chaotic Scaling." Until then, keep your transactions off-chain and your memes spicy.

Teaser for Chapter 7:

"Next time: Blockchain isn't just for degenerates! Discover how it's saving avocados, curing hospitals of fax machines, and why your vote might soon live on a blockchain (unless the guy in the Capitol basement still loves Excel)."

CHAPTER SEVEN

BLOCKCHAIN IN ACTION – REAL-WORLD USE CASES

Finance and Banking: The Digital Bank Heist (Without the Cops)

Gone are the days when sending money across borders involved a carrier pigeon and a prayer. Blockchain's brought cross-border payments into the 21st century like a heist movie where everyone gets away clean.

Cross-Border Payments:

- Remember sending money overseas and praying it doesn't end up in a black hole? Blockchain flips that on its head. No more crazy fees or waiting for days. With blockchain, sending money

across the globe is like sending an email—fast and cheap.
- Example: If you're in India trying to send money to a friend in Mexico, forget Western Union. With blockchain, you'll both get your cash faster than your friend can say "DeFi."

Decentralized Finance (DeFi):

- DeFi is the new frontier where financial services work without any banks, middlemen, or that one uncle who swears he can predict stocks. Instead, you can lend, borrow, trade, or just stake your crypto and make money, all in the wild west of finance.
- Example: Imagine borrowing funds for a new venture but not needing to beg a bank manager for approval. DeFi makes that possible—minus the suits and paperwork.

Pro Tip: DeFi's like playing Monopoly, but instead of a paper banker, there's a smart contract. No more cheating, and everyone wins—except the old-school financial system.

Supply Chain: From Mangoes to Military Gear – Blockchain's Tracking Superpower

Blockchain's got an obsession with proving things are legit—like that you really didn't eat the last slice of pizza.

Tracking Goods from Source to Consumer:

-

Whether it's a pair of Gucci sneakers or a mango from Mexico, blockchain makes sure you know exactly where it came from and whether it's real or just a knockoff.

- Example: Walmart uses blockchain to track food safety. Imagine knowing exactly where your lettuce came from and how many people touched it before it ended up on your plate. (Not to mention, you can check if that mango's a direct import from an organic farm or just some supermarket lie).

Pro Tip: If you're a supply chain manager using blockchain, you're basically the detective in an international thriller, minus the trench coat and the cigar.

Healthcare: The Digital Health Revolution (Finally! No More Faxes!)

-

If your medical records still look like they've been copied on a fax machine from the 90s, blockchain's here to change that.

-
- Securing Medical Records:
- Blockchain ensures your health data is encrypted, secure, and most importantly, doesn't get lost in some clerk's pile of paperwork. Medical professionals can access your history instantly, with a few clicks—not a few weeks.
- Example: Estonia, that tiny country with more e-governance than a Silicon Valley startup, uses blockchain to store all medical records. Yes, even your grandma's flu shot is on blockchain.

Pro Tip: Think of blockchain in healthcare as your super-secure digital doctor who remembers everything—without the awkward small talk.

Voting and Governance: The Polls Get Real (And Real Secure)

- Remember those elections where people weren't sure if their vote counted? Blockchain's here to throw those concerns out the window. Welcome to tamper-proof elections.
- Transparent and Tamper-Proof Elections:
- Blockchain's decentralized nature makes it perfect for voting. No more rigged machines or hanging chads. With blockchain, you vote, and it's recorded exactly as you intended. No funny business.
- Example: Sierra Leone's blockchain-based election in 2018 was the start of a new era. The election results were verified transparently and quickly, without anyone questioning the integrity of the process.

Pro Tip: If elections were run on blockchain, we'd all probably be way more excited about democracy. No more debates about who's cheating and who's not.

Art and NFTs: Digital Art Meets the Blockchain Mafia

- *You thought NFTs were just for bored billionaires and weird pixelated apes? Think again. Blockchain's making art ownership a digital reality—and a very lucrative one at that.*

- Digital Ownership and Provenance:
- With NFTs (Non-Fungible Tokens), artists can prove they own their work digitally—and buyers can own a piece of it. No more "is this a fake?" dilemmas, thanks to the blockchain proving who's the rightful owner.
- Example: Beeple's infamous $69 million NFT sale is proof that blockchain isn't just for crypto bros. It's for art collectors who like to hang JPEGs on their digital walls.

Pro Tip: NFTs are like owning a digital Mona Lisa, except instead of hanging it in your living room, you've got it stored on the blockchain—and it's probably worth more than your house.

What's Next?

In Chapter 8, we'll dive deep into the futuristic world of blockchain, exploring its potential to reshape industries and lives. From the decentralized internet revolution (Web3) to the quantum leap in computing, we'll break down where blockchain is headed—and how it's preparing for the biggest technological challenge it has ever faced.

Find out how interoperability will change the way different blockchains talk to each other, why your data might soon be in the hands of decentralized apps (goodbye, Google?), and how blockchain's war with quantum computing could redefine everything we know about security. The future is bright, decentralized, and possibly a little bit terrifying.

Glossary: Speak Like a Blockchain Pro at Your Next Crypto Party

Cross-Border Payments: Sending money across borders without the hassle of banks, fees, or waiting. Think of it like transferring files between your phone and computer, except it's cash.

DeFi (Decentralized Finance): A digital financial ecosystem without banks or intermediaries. If you've ever wanted to lend money to a stranger and trust that they'll pay you back with no paperwork—this is it.

NFT (Non-Fungible Token): A unique digital asset that proves ownership of a digital item. It's the certificate of authenticity for anything you can imagine, from art to tweets.

Smart Contracts: Self-executing contracts with the terms directly written into code. They work like a vending machine—insert the right stuff, and out comes the product.

Sierra Leone's Blockchain Election: A real-world example of using blockchain for elections. No hanging chads here, just transparent, verifiable votes.

Provenance: The verified history of an item, proving it's authentic—like knowing the artist behind that $69 million NFT.

Teaser for Chapter 8:

"Next time: Hold onto your seats, because we're about to take a wild ride into the future of blockchain. Will it survive quantum computing? Can Web3 become the next big thing, or is it just a bunch of decentralized pipe dreams? Find out in Chapter 8, where we predict the world of tomorrow—powered by blockchain, of course!"

CHAPTER EIGHT

The Future of Blockchain Trends and Predictions

Web3 and the Decentralized Internet: The Web You Own (No, Really, You Do)

Alright, buckle up because we're entering the new frontier: Web3. This is the decentralized future where your data is yours, and you get to decide who plays with it. Forget about those mega-corporations that hoard your data like Gollum with his precious. In Web3, you're the boss. (And you don't even have to pay in likes.)

What is Web3?

- Imagine a world where every app you use doesn't store your personal data in some giant corporate warehouse. In Web3, the internet is owned by you, the users, through decentralized apps (dApps). Instead of your data being a product to sell, you're in control of it—like having the keys to your digital house and kicking out any uninvited guests.

- Web3 is built on blockchain, which means it's trustless, transparent, and resistant to censorship. It's like the wild west of the internet, only with better security and less cowboys trying to sell you fake watches.

Web2 vs. Web3: A Comparison

- Web2: Your social media accounts are owned by Mark Zuckerberg (and he's definitely looking at your vacation photos). Every time you search something, Big Tech is lurking behind the scenes, compiling your data and selling it for profit.
- Web3: You own your data. Instead of handing over your soul for a free app, you pay with cryptocurrency or even better, you trade value for value. Everything is decentralized—think of it as the opposite of being trapped in a Facebook black hole.

Pro Tip: Web3 is like the superhero of the internet, and Web2 is the villain with a secret lair in Silicon Valley. Choose your side wisely.

Interoperability: When Blockchains Start Speaking to Each Other Like Old Friends

For too long, blockchains have been like those annoying relatives who won't talk to each other at family gatherings. Polkadot and Cosmos are like the digital peacekeepers, working hard to make sure everyone can get along. They're the matchmaking service for blockchains—here to make sure no chain feels isolated or misunderstood.

Why Blockchains Need to Communicate

- Imagine you're at a party and you can't talk to anyone because they only speak their own language. That's what it's been like in the blockchain world. But with interoperability, different blockchains can finally understand each other, exchange data, and work together—just like a team of superheroes combining their powers.

- Polkadot is working to create a web of blockchains that can transfer information seamlessly. It's like creating a superhighway for blockchains to share traffic without causing a pile-up. Cosmos is also doing its part, building a network that allows independent blockchains to interact without getting all tangled up.

Pro Tip: Interoperability is the blockchain equivalent of introducing your weird cousin to your cool friend and watching them get along. It's magic, it's chaos, and it's going to change everything.

Regulation and Adoption: The Blockchain Big Brother You Don't Have to Fear (Maybe)

Governments are catching up. Finally. It's like that time when your mom figured out how to use the internet, and you were like, "Whoa, this is happening." Blockchain regulation is coming, and while we don't know exactly what it will look like, we do know that it's going to have a huge impact on how blockchain gets adopted.

How Governments Are Approaching Blockchain

Some countries are like the cool uncles who just let blockchain do its thing (looking at you, El Salvador and their love for Bitcoin). Others are more like the overprotective parents (we're talking to you, China, with your digital yuan).

How Governments Are Approaching Blockchain

- China's Digital Yuan: China's not just dipping its toes into blockchain; they're jumping in headfirst with a central bank digital currency (CBDC). The digital yuan will be fully controlled by the government, but it's a huge step toward embracing blockchain technology in the public sector.

- El Salvador's Bitcoin Adoption: Meanwhile, El Salvador is going rogue, making Bitcoin legal tender. It's the rebellious teenager of the blockchain world, and honestly, we're kind of here for it. While the world holds its breath to see how this experiment pans out, El Salvador might just be on to something.

Pro Tip: Blockchain is like a rebellious teenager trying to figure out what to do with its future. Some countries are sending it to college; others are threatening to ground it forever. We're all just waiting for the graduation ceremony.

Quantum Computing and Blockchain: The Apocalypse or Just Another Tuesday?

Ah, quantum computing—the shiny, mysterious tech that promises to break everything we thought we knew about security. So, what happens if quantum computers are able to break through blockchain's cryptographic defenses? Panic? Nah. Blockchain's already thinking about it.

The Threat of Quantum Computing

- Quantum computers can theoretically solve problems that would take a classical computer millennia, and that includes cracking cryptographic codes used by blockchain. This is like a giant monster lurking in the digital ocean, threatening to smash through the walls of blockchain security.
- But don't worry. Blockchain has a backup plan. Many blockchain projects are already working on "quantum-resistant" algorithms to keep those pesky quantum computers from ruining the fun. It's like having a bodyguard who's already preparing for the zombie apocalypse—even though it hasn't happened yet.

How Blockchain is Preparing for the Quantum Era

Think of it like preparing for a hurricane. You don't know when it'll hit, but you know you've got to make sure your house is stormproof. Blockchain's got its quantum-proofing plans in place, and while quantum computers might still be a few years away from becoming a real threat, the community is already ahead of the game.

Pro Tip: Quantum computing is like the plot twist in a sci-fi movie. It's big, it's scary, but blockchain's got the cool gadgets and a plan to save the day.

Final Thoughts: Blockchain's Potential – More Than Just a Fad

So here we are at the end of our blockchain journey. You've learned about Layer 1 and Layer 2, you've explored real-world use cases, and you've even had a few laughs along the way. Now, as we stand at the edge of the blockchain revolution, it's time to reflect on just how transformative this technology really is.

Blockchain: The Ultimate Disruptor

Blockchain's not just a trend; it's a seismic shift in how we handle data, transactions, and trust. It's the start of a new digital era, where ownership, privacy, and control are put back into the hands of individuals. Whether you're a finance geek, a supply chain guru, or a digital artist, blockchain is something you're going to want to keep an eye on.

It's the technology that can take down corporations, streamline industries, and possibly change how we govern ourselves. It's not just about Bitcoin or Ethereum—it's about reshaping the world itself.

Pro Tip: Blockchain is the future. It's like the internet in the 90s—confusing, wild, and full of possibilities. Get on board, or risk being that guy who thought the internet was just a passing phase.

What's Next?

You've made it through the wild, weird world of blockchain, and now it's time for you to take your knowledge to the next level. Whether you're looking to dive deeper into smart contracts, build your own decentralized app, or maybe even start your own blockchain project, the world is your oyster. (And it's a decentralized oyster, by the way.) Get involved, explore, and who knows—you might just end up creating the next big thing.

Glossary: The Blockchain Dictionary You'll Actually Use

- Web3: The decentralized internet, where you own your data and everyone shares control.
- Interoperability: The ability of different blockchains to work together. Think of it as the universal translator for crypto.
- CBDC (Central Bank Digital Currency): A digital currency issued by a country's central bank. Basically, the government's version of Bitcoin—just less fun.
- Quantum Computing: The brain-melting technology that can crack current cryptographic systems.
- Quantum-Resistant Blockchain: Blockchain that's prepared for the quantum computing apocalypse.
- DeFi (Decentralized Finance): Financial services without middlemen or banks—just smart contracts and peer-to-peer transactions.
- NFT (Non-Fungible Token): A digital item proving ownership of a unique asset, often used for art, collectibles, and the occasional cat meme.

- Polkadot/Cosmos: Blockchains that want to talk to each other, like a digital social network for crypto.

Teaser for the Sequel (Maybe?)

"What's next in the world of blockchain? We're not sure, but we're excited to find out. Join us as we explore how blockchain will tackle the next big challenge: maybe solving world hunger, maybe building a decentralized version of Netflix, or maybe creating the first ever blockchain-powered virtual pet rock."

—powered by blockchain, of course!"